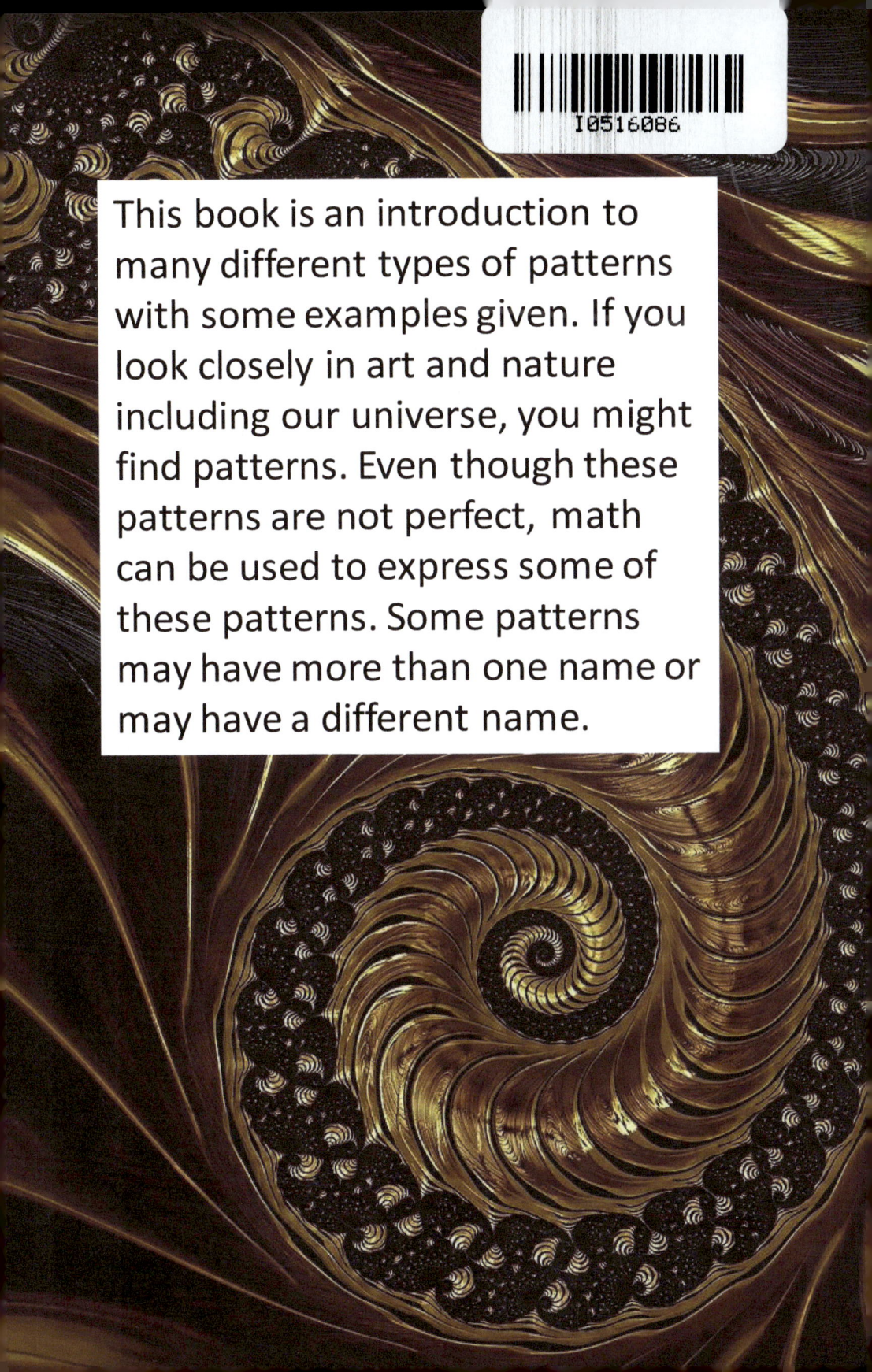

This book is an introduction to many different types of patterns with some examples given. If you look closely in art and nature including our universe, you might find patterns. Even though these patterns are not perfect, math can be used to express some of these patterns. Some patterns may have more than one name or may have a different name.

A pattern is a repeated design that is found everywhere in our world and even in outer space.

All around you there are both simple patterns and complicated patterns. The spiral of a seashell is very complex.
But, there are patterns more complicated than a seashell.

When did humans discover or invent patterns? There are cave paintings that are around 10,000 years old. They show hands in a pattern with 5 fingers.

For ancient people, their fascination with their hands probably helped them invent numbers and discover patterns for other objects around them like counting their goats or measuring the tallness of trees to cut down and build shelters.

Over 4,000 years ago, ancient Egyptians used geometry which is the study of shapes and patterns. This knowledge allowed them to build objects like the step pyramid found in the Egyptian desert.

Around 1200 AD, Fibonacci (fee-buh-NAW-chee), an Italian mathematician, wrote about a problem using an idealized population growth of rabbits. What do you think that sequence was like?

Start with adding 0 and 1 then adding 1 and 1. The next number is the sum of the previous two. Repeat this pattern and you'll get this series to infinity: 0,1,2,3,5,8,13,21,34,55,89,144, ... After only a few off-springs you would ideally have 144 rabbits!

A variety of plants like the head of the yellow chamomile flower have consecutive Fibonacci numbers in their spiral arrangement. Where else might Fibonacci numbers be found?

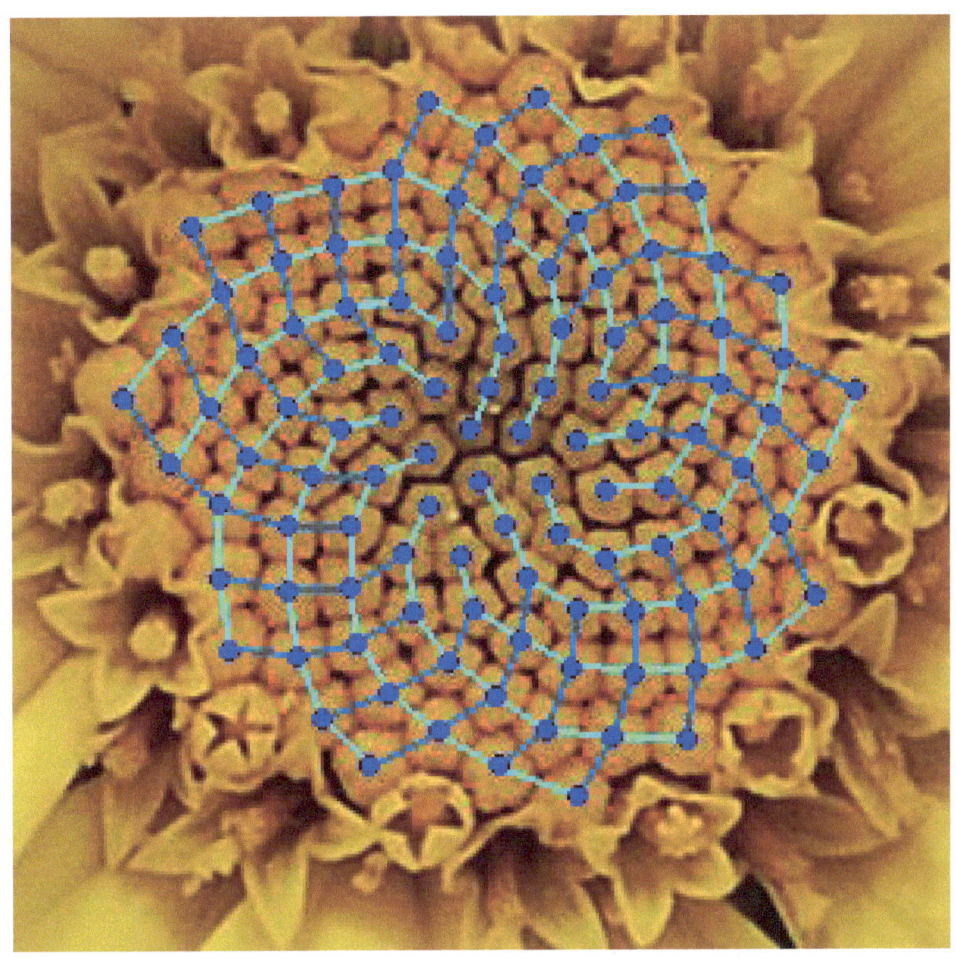

Petals in many flowers show Fibonacci numbers. Lily flowers have 3 petals, buttercups have 5, white dryads have 8, Indian summers have 13, chicory flowers have 21, daisy flowers can have 34 or 55 petals. Why are these Fibonacci numbers in flowers?

Fibonacci numbers (0,1,2,3,5, etc.) are also found in tree branches, as well as ferns, pineapples, artichokes, pinecones, and the security codings of cellphones and computers.

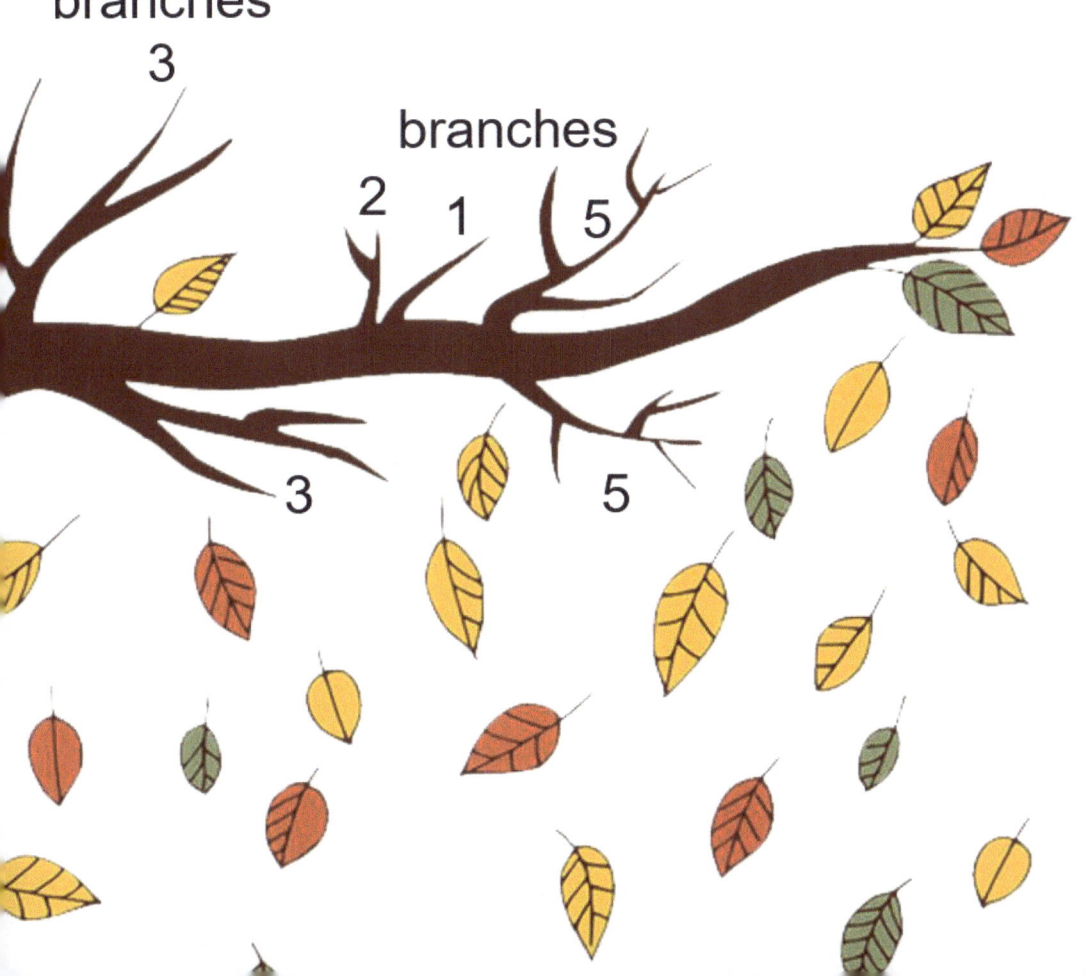

The reason why the Fibonacci numbers are found in tree branches, plant seeds and flower petals is for each plant to make use of the best space possible and to grow in the best way possible.

In 1904, Helge von Koch (hel-GUH von-KAWCH) wrote about a never-ending pattern called a fractal. The Koch snowflake is a fractal curve. To draw this pattern, you start with a triangle. Then you cut out the middle third and draw another triangle. You repeat this method until you have what looks like a snowflake.

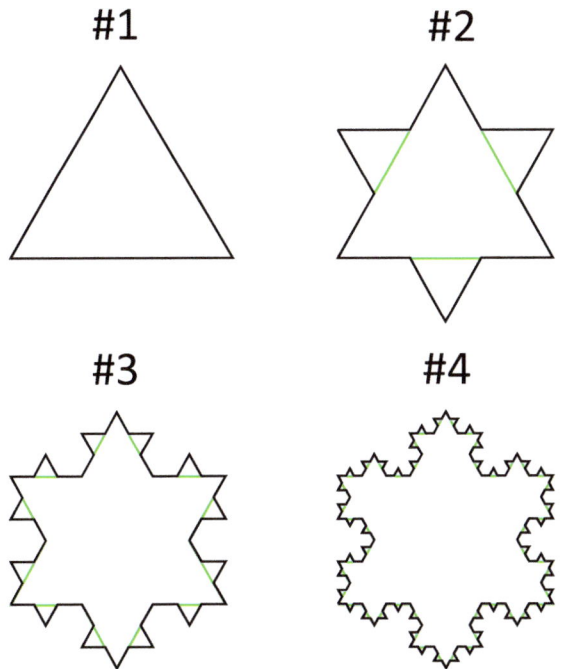

Some real snowflakes in nature with their repeating pattern seem to look like the mathematical Koch snowflake. It appears that nature can be a mathematician.

The Golden Spiral is a special spiral. It has an unending number with a value of 1.618...
Where in our universe do you think that the Golden Spiral is found?

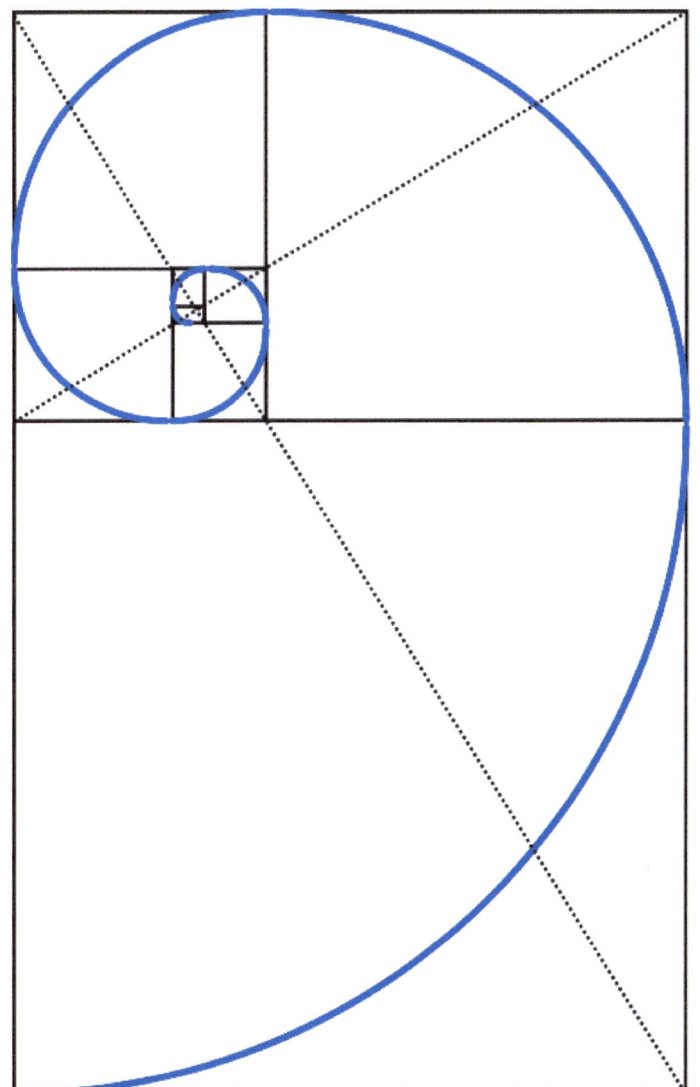

Among other places, the Golden Spiral was discovered in spinning, spiral galaxies like M51 which has billions of stars in two arms slinging out from its center.

Humans and other animals are made up of fractals. Our brains, lungs, and circulatory system are like fractal trees. Many natural objects are made up of many different types of fractals woven into each other.

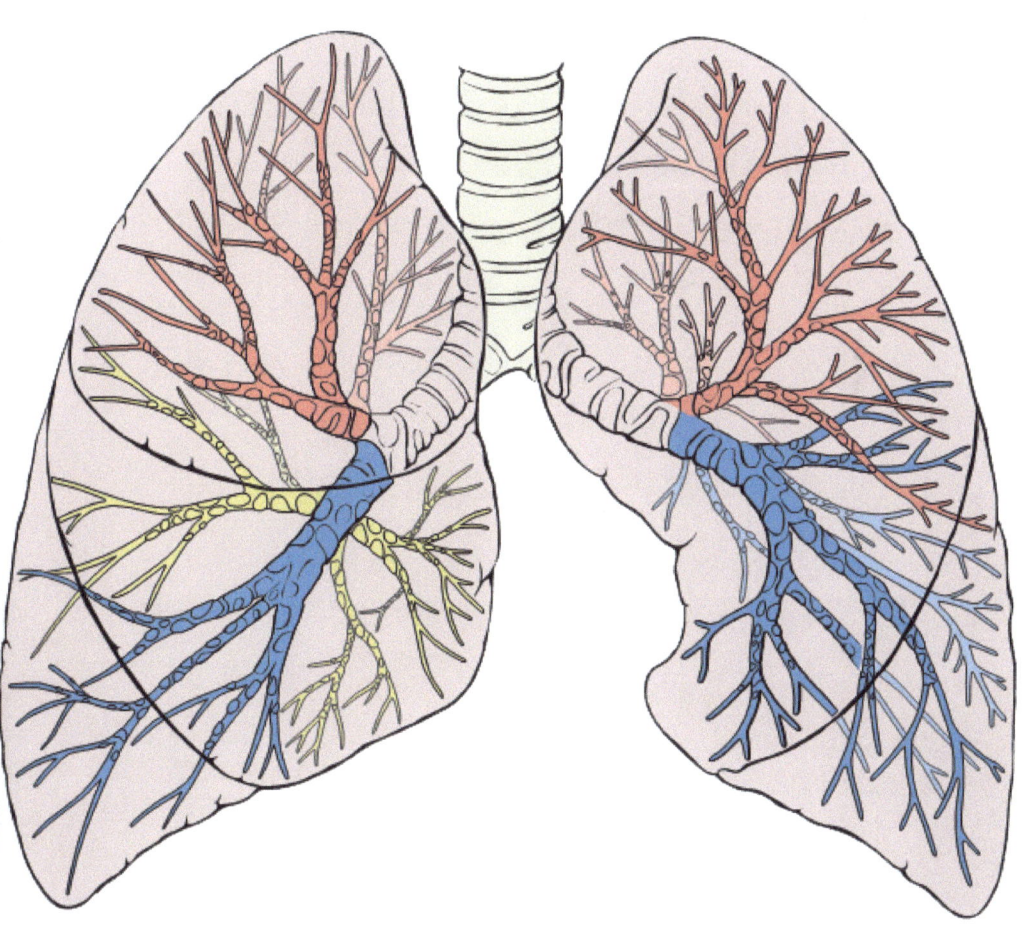

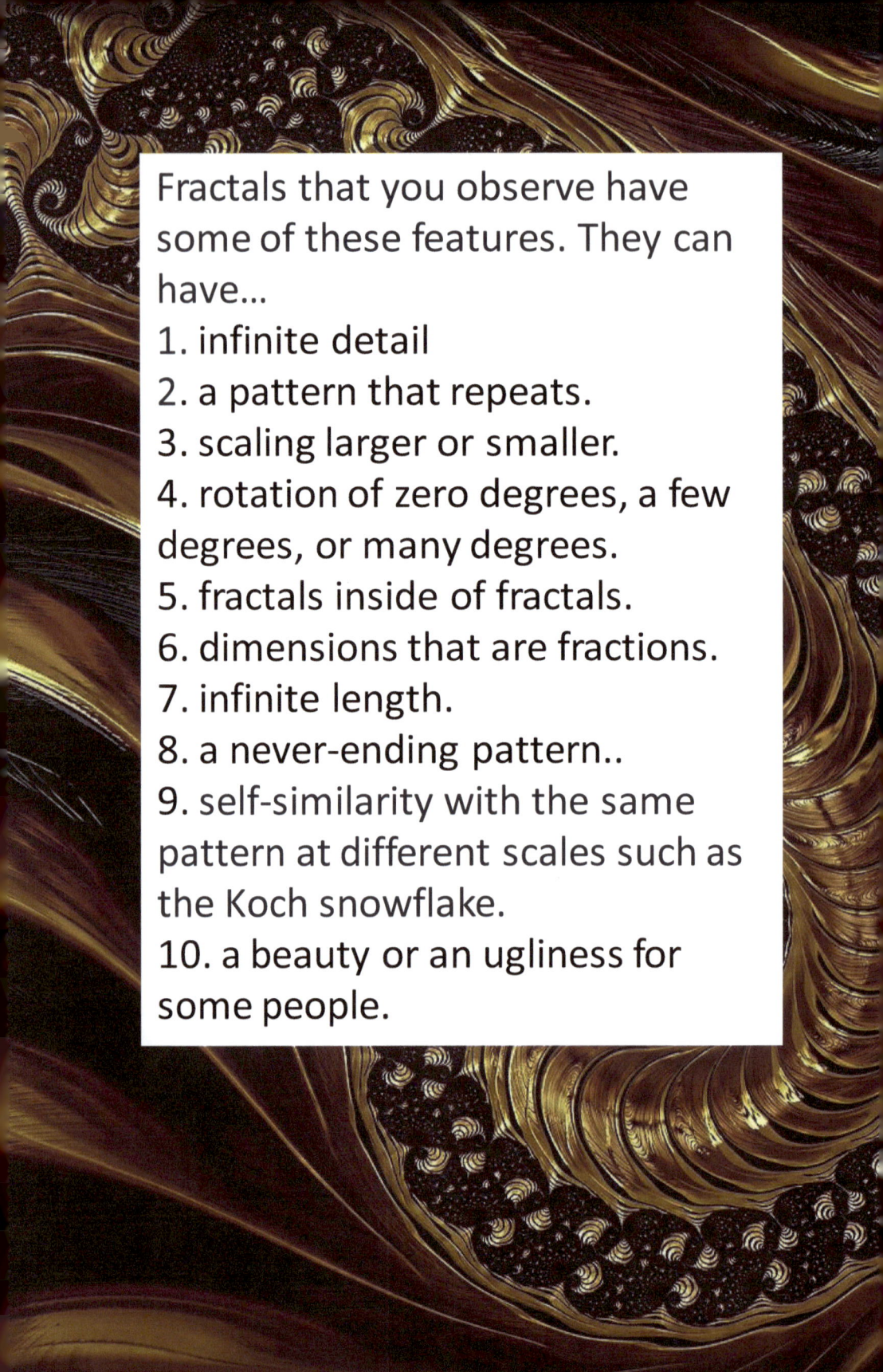

Fractals that you observe have some of these features. They can have...
1. infinite detail
2. a pattern that repeats.
3. scaling larger or smaller.
4. rotation of zero degrees, a few degrees, or many degrees.
5. fractals inside of fractals.
6. dimensions that are fractions.
7. infinite length.
8. a never-ending pattern..
9. self-similarity with the same pattern at different scales such as the Koch snowflake.
10. a beauty or an ugliness for some people.

A fractal heart is made up of smaller, rotated, stretched hearts.

A tessellation (teh-suh-LAY-shun) is an arrangement of shapes closely fitted together. There are three types of tessellations: translation (sliding), rotation (spinning), and reflection (flipping).

Tessellations are observed everywhere in our world and outer space.

Early people saw tessellations in nature.
The honeycomb of bees is a natural tessellated structure of six-sided shapes.

Around 3,000 BC, the Sumerians used colored tiles and arranged them in a tessellation pattern.

The word "tessellar" comes from an Ancient Greek word meaning "four" and refers to the term "tiling" that often uses fire glazed clay. The tiles were used to decorate walls, ceilings, and floors.

In Latin, the word "tessellar" means a small cubical piece of glass, stone, or clay used to make mosaics. A mosaic is a picture or pattern produced by arranging together small colored pieces of a hard material.

Some flat Roman mosaics had a 3-dimensional look.

In Islamic art, tessellations were used in the decorative tiling of the Alhambra palace.

Tessellations would inspire M.C. Escher (ESS-sure) in some of his artwork like his tessellation of a fish and a bird or of two flying birds.

Johannes Kepler (Yo-HAN-es KEP-ler) made a study of tessellations in 1619. Writing about regular and semiregular tessellations, he was possibly the first to explore and to explain the six-sided structures of bee honeycomb and snowflakes.

In 1891, the work of the Russian Fyodorov (FEE-oh-door-off) marked the unofficial beginning of the mathematical study of tessellations. He wrote "Basics of Polytopes".

A polytope is a geometric object with flat sides. Two-dimensional polytopes are called polygons and three-dimensional polytopes are called polyhedra.

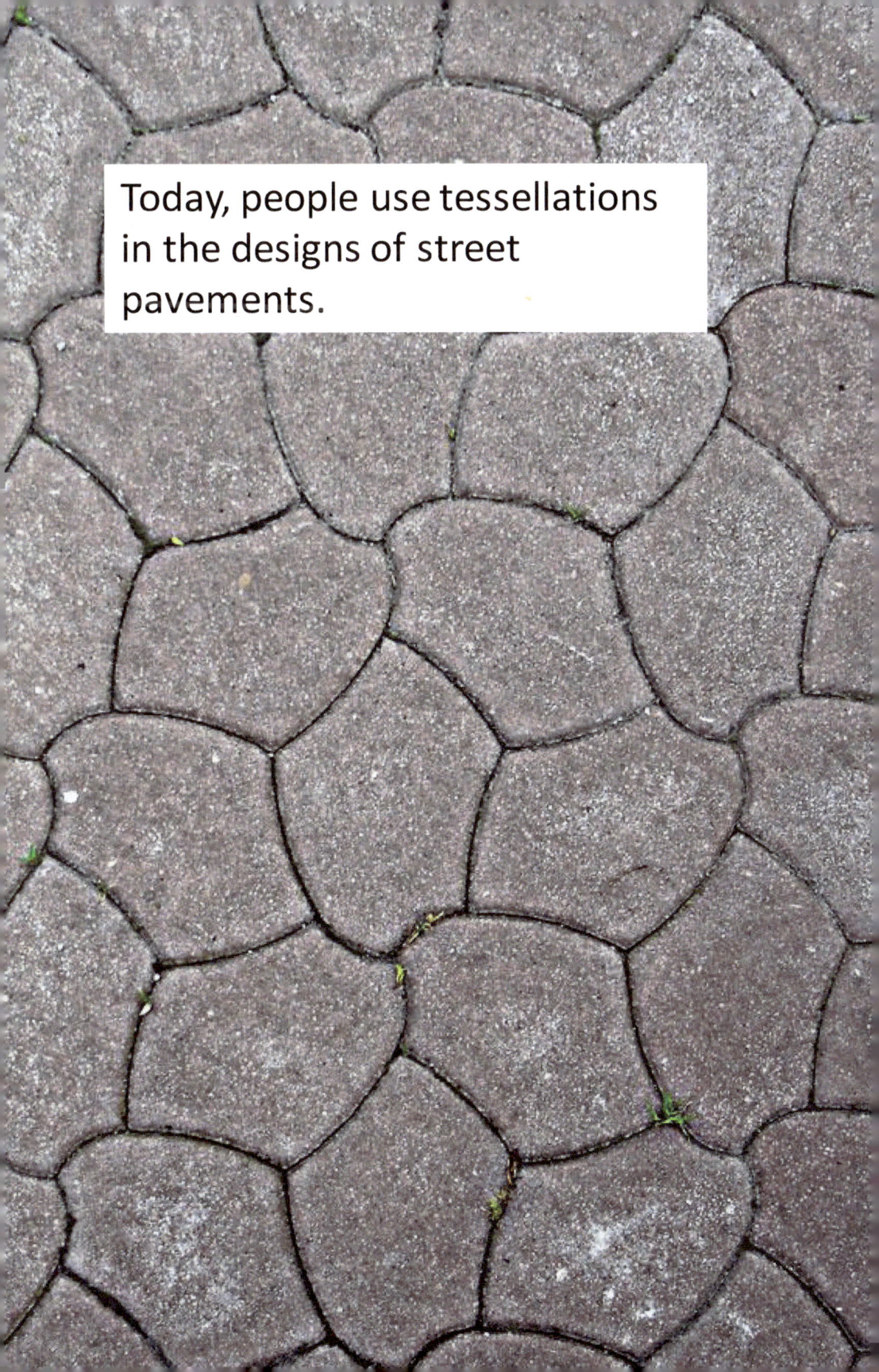
Today, people use tessellations in the designs of street pavements.

Tessellations can be used in floor covering.

Penrose tilings use two different four-sided tiles. These tiles forcibly create non-periodic patterns.

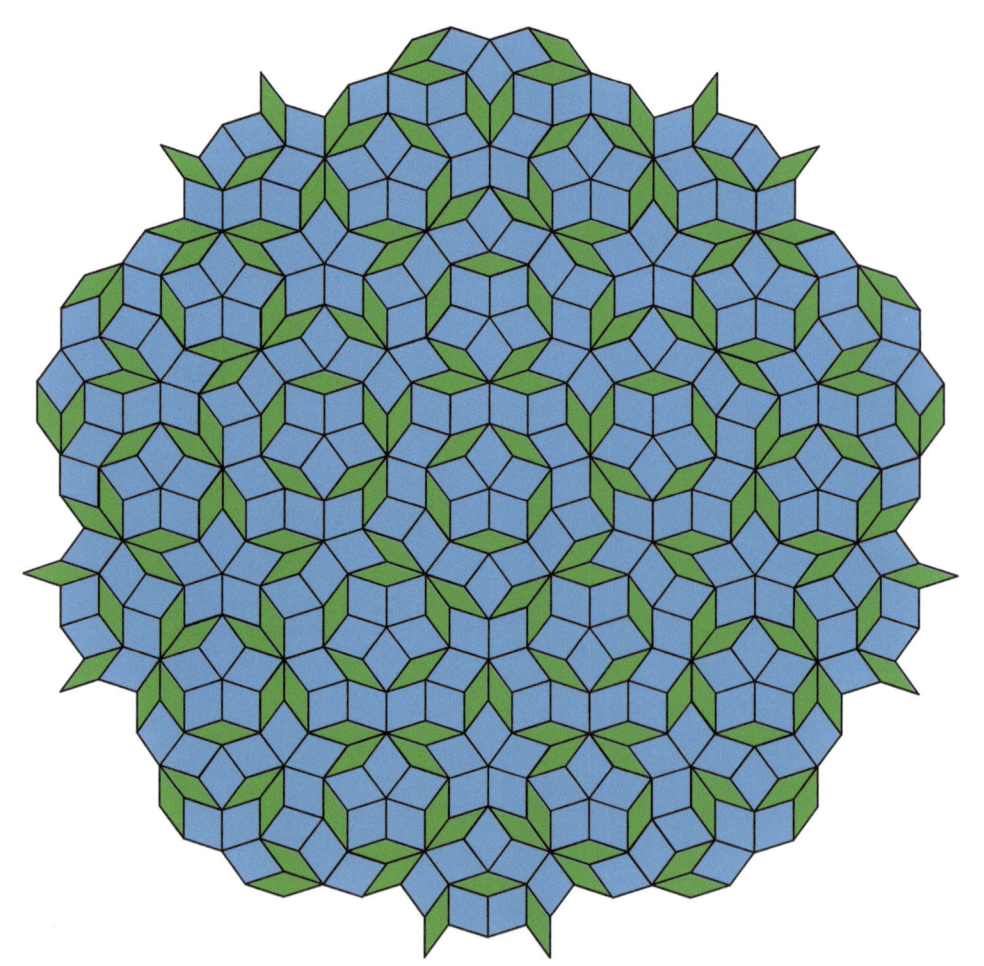

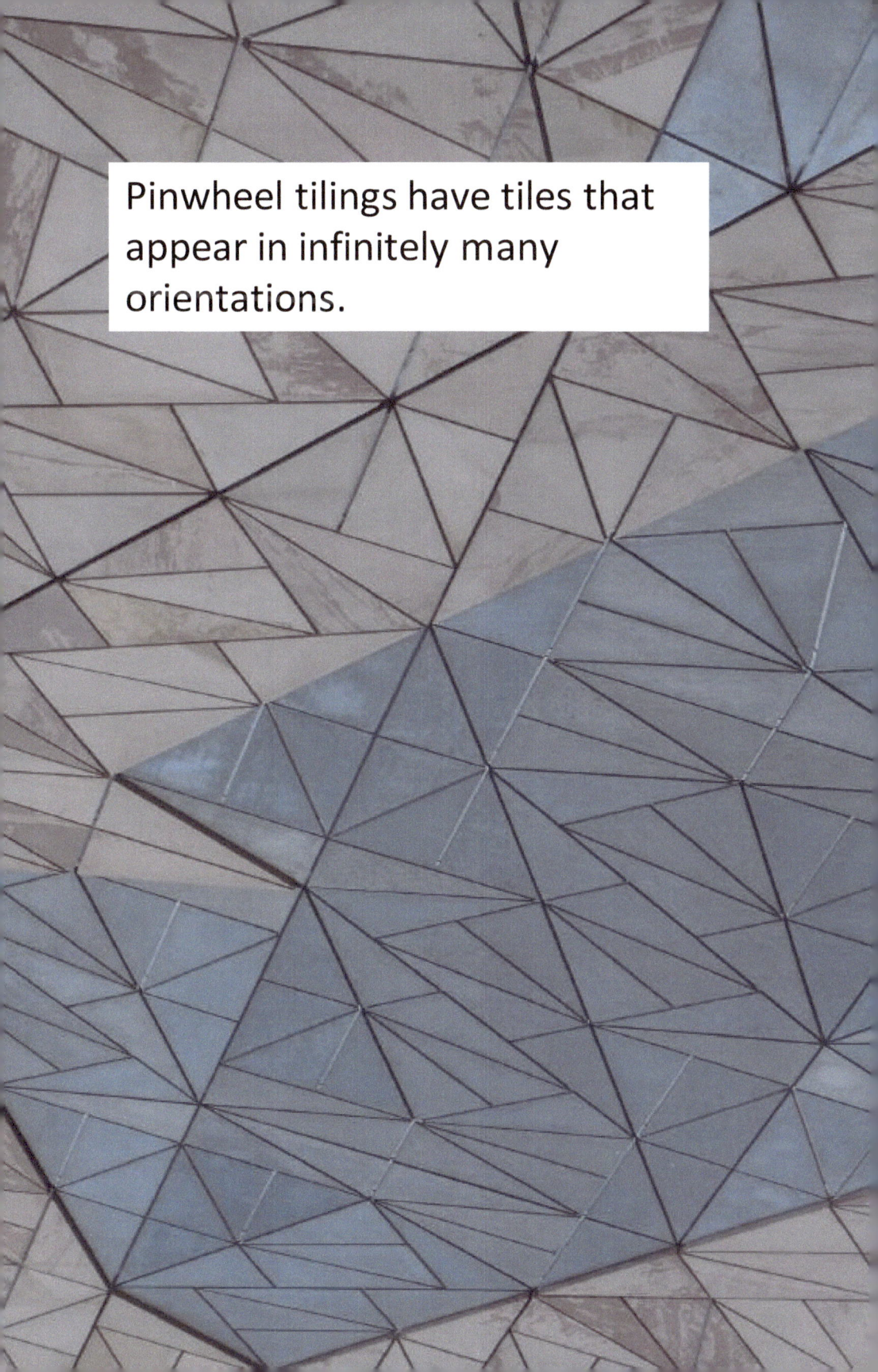

Pinwheel tilings have tiles that appear in infinitely many orientations.

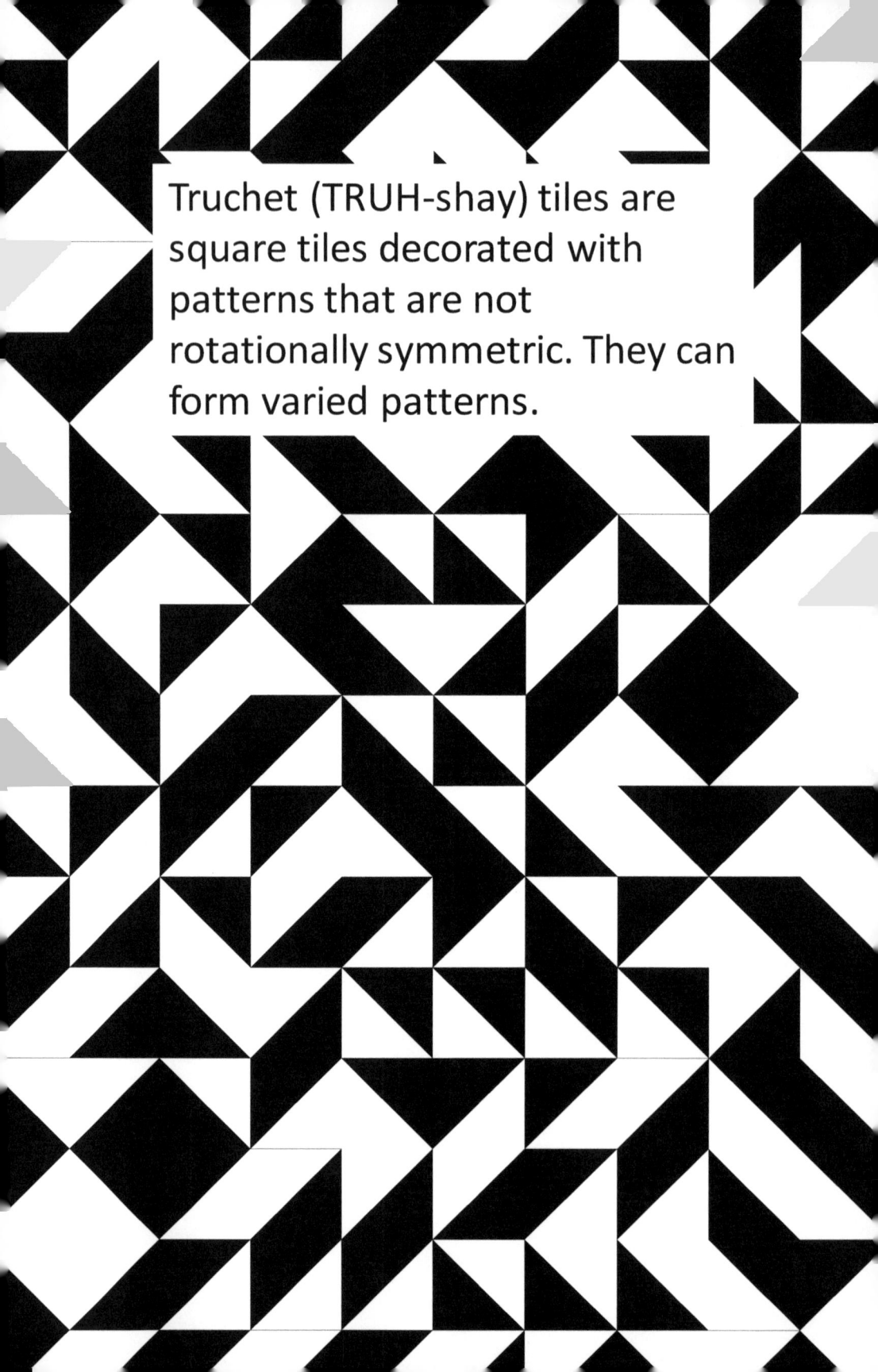

Truchet (TRUH-shay) tiles are square tiles decorated with patterns that are not rotationally symmetric. They can form varied patterns.

Voronoi (VOR-uh-noy) tilings have applications in almost all areas of science and engineering. Structures of living things can be described using them. They can help identify the nearest airport in case of diversions in aviation. They also aid estimation of overall mineral resources based on exploratory drill holes in mining.

Tessellation can be made in three dimensions. Some can be stacked in a regular crystal pattern.

Some 3-D tessellations can be found in nature such as crystals of a fluorite mineral.

Some flowers like the autumn crocus have a tessellate pattern in their petals.

Saturn is surrounded by over 1000 rings made of ice and dust. There are 7 main large rings. The ice and dust in the rings of the planet Saturn form a tessellation in the form of concentric rings that are smaller rings inside of larger rings with the planet as the center.

Tessellations are used in puzzles and recreational mathematics like the Tangram dissection puzzle.

The triangular tiling alternates the color of one triangle with the color of an adjacent triangle

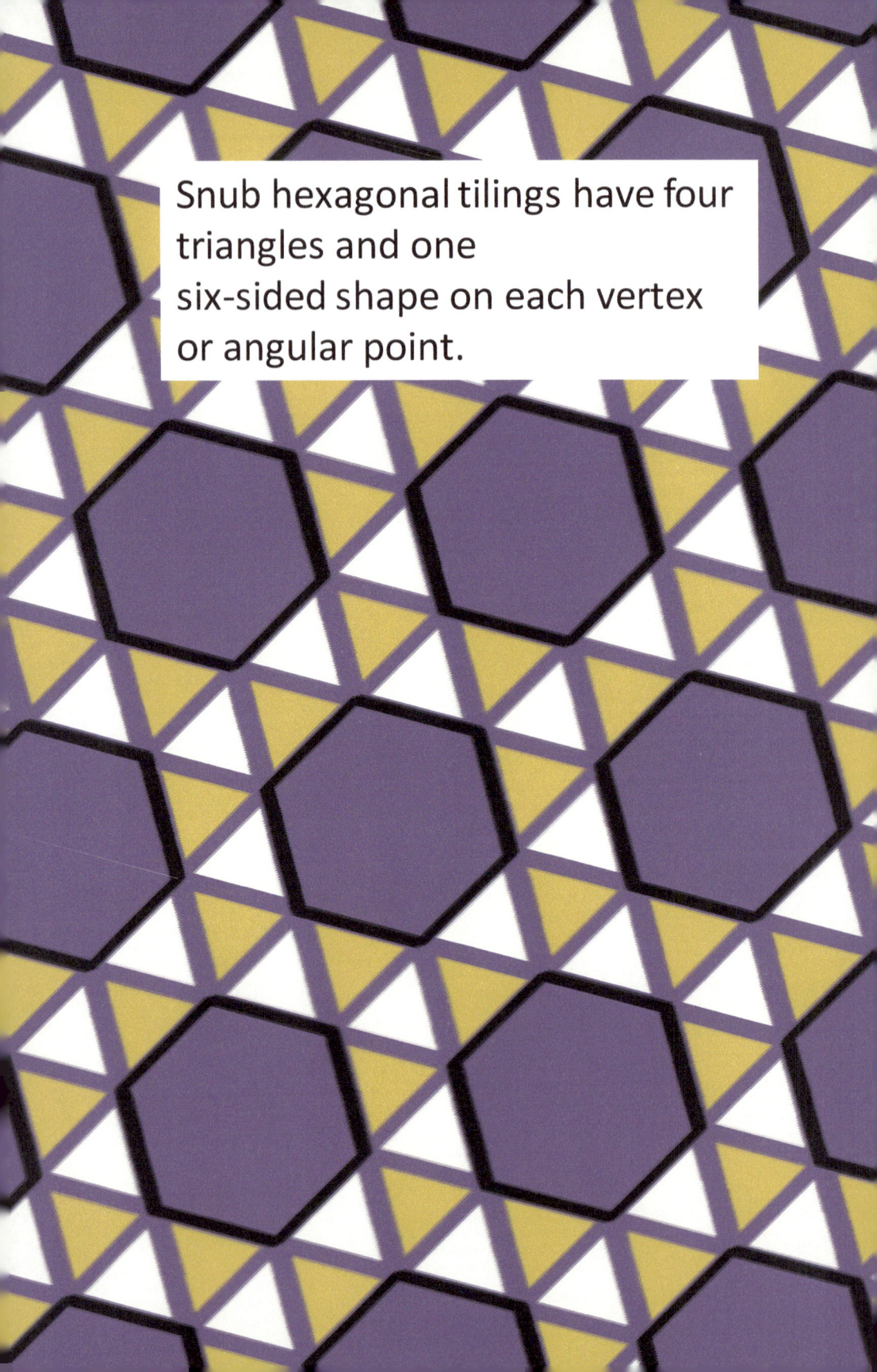

Snub hexagonal tilings have four triangles and one six-sided shape on each vertex or angular point.

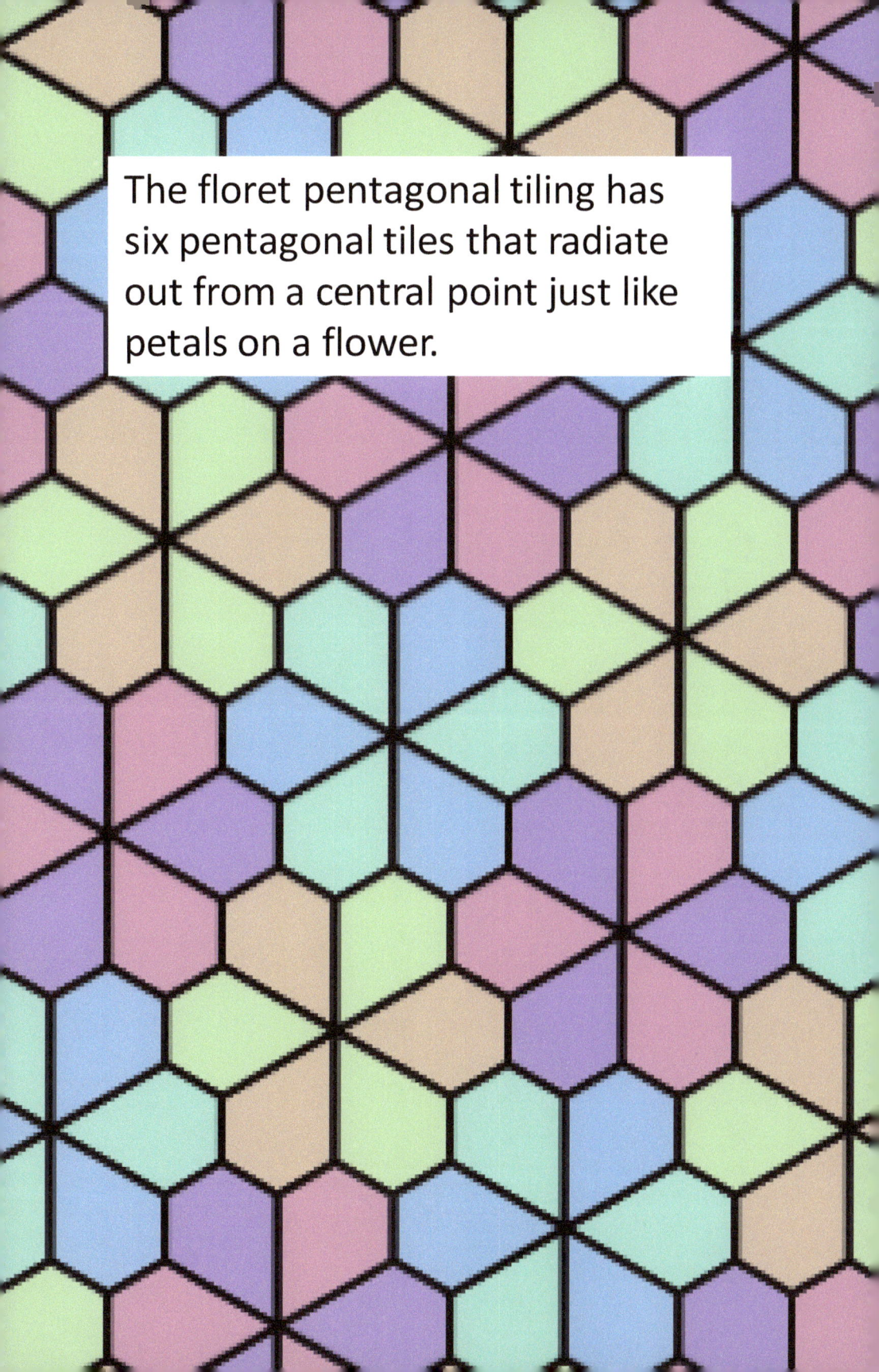

The floret pentagonal tiling has six pentagonal tiles that radiate out from a central point just like petals on a flower.

The Voderberg tiling is a spiral tiling made of nonagons or nine-sided shapes.

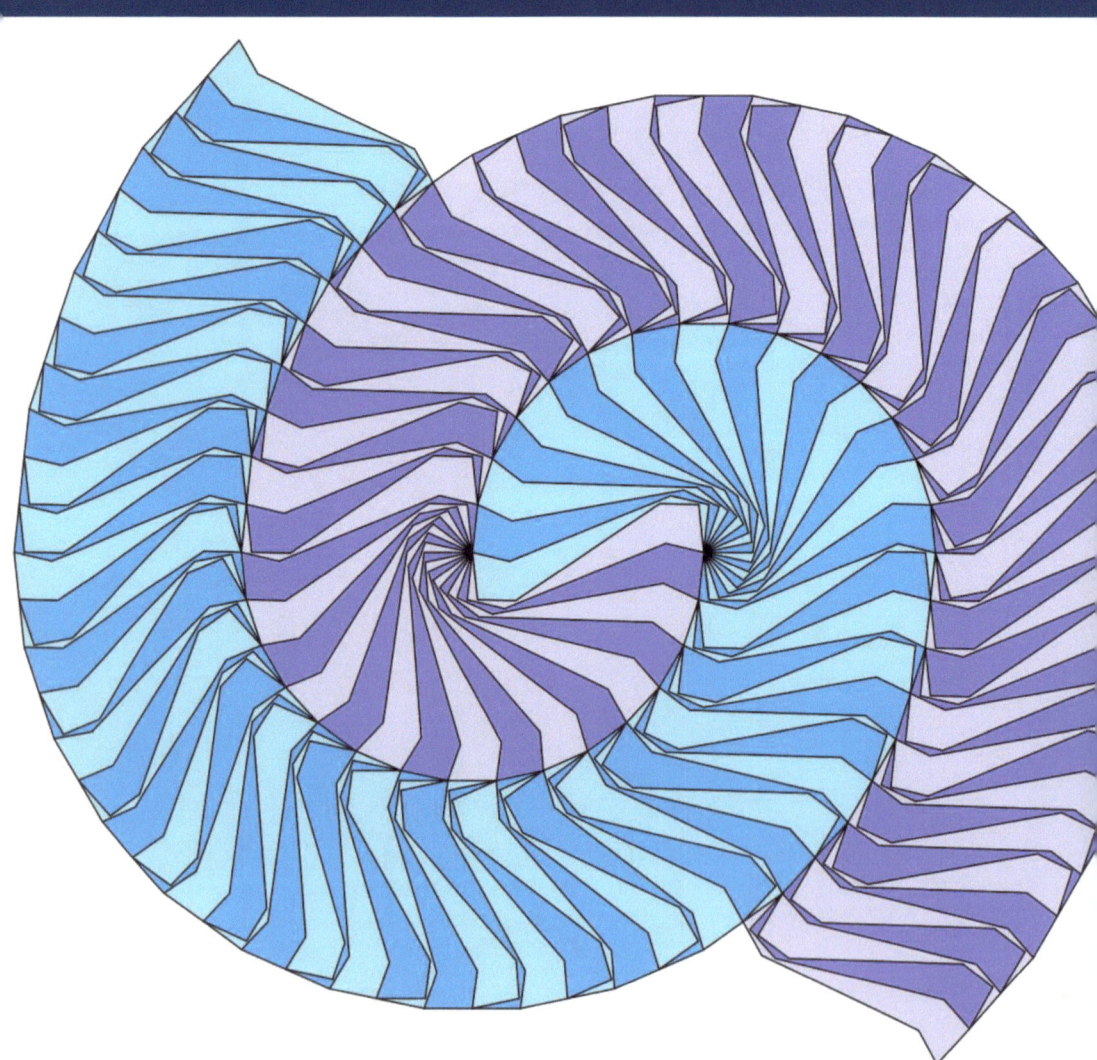

The tritetragonal (TRY-tuh-trag-uh-nole) tiling is a tessellation that involves the geometry of saddle surfaces.

A tessellation can be made using the capital letter I and rotating it.

There are some origami constructions that are tessellations.

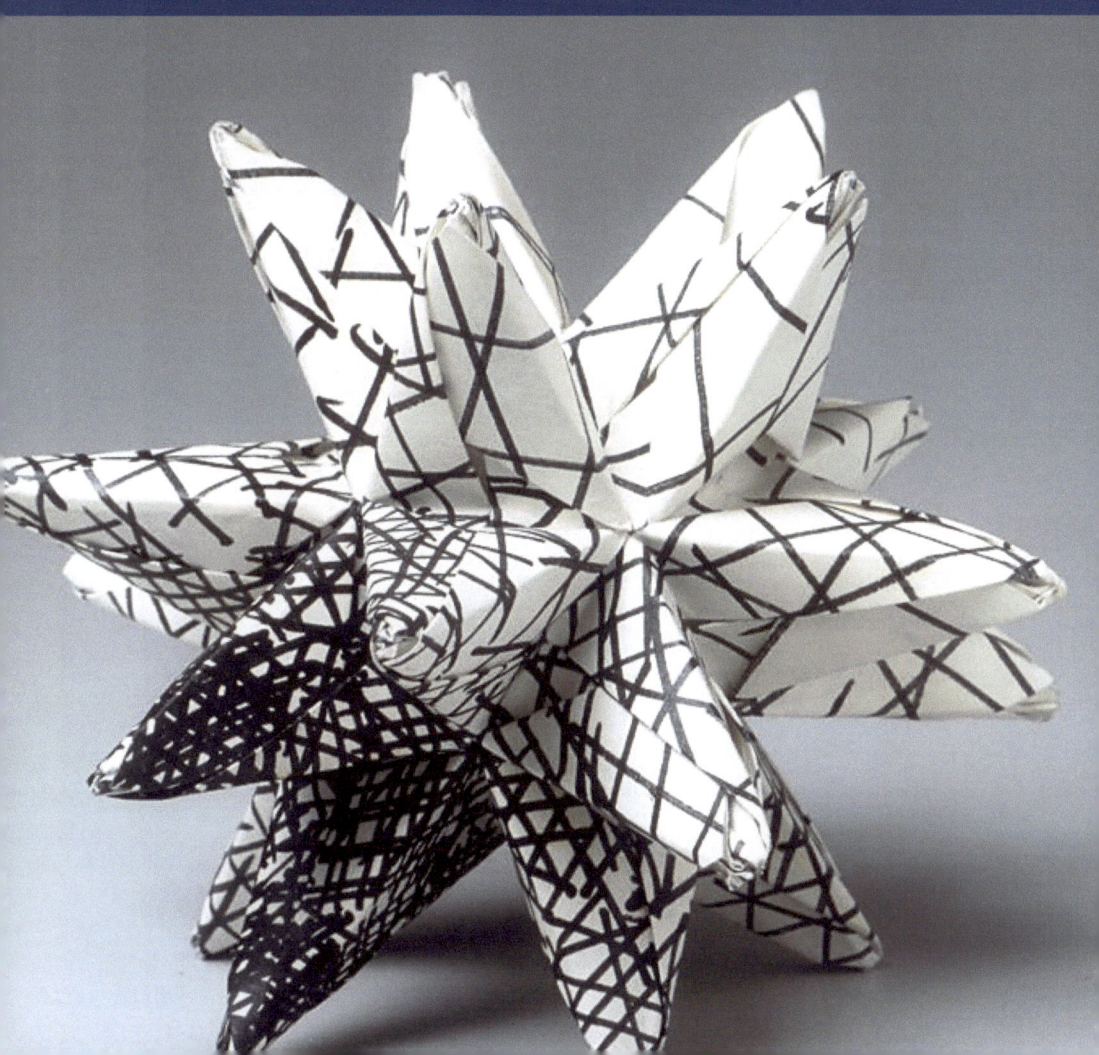

There are many examples of tessellations found around us in nature, outer space, hobbies, architecture, and the arts. How do you see or use tessellations in your life?

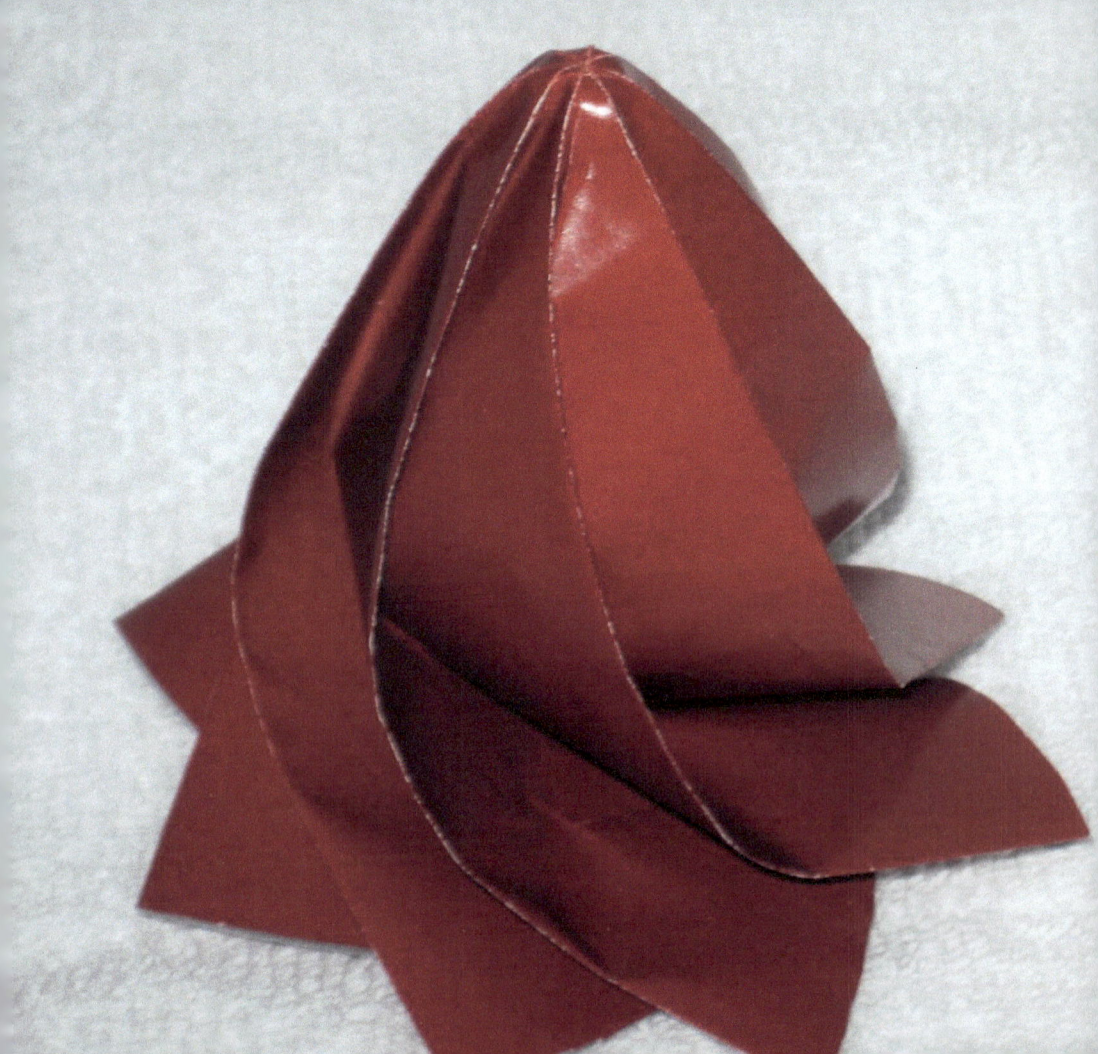

"Do not worry about your difficulties in mathematics, I assure you that mine are greater."
Albert Einstein

Prime Numbers between 1 and 100
An Easy Way to Find Them

The word prime itself means most important or of first importance.

What do you think prime numbers are?

Prime numbers are certain counting numbers (natural numbers or positive integers) that start with the positive number 2 and go on.

" I am zero and I'm tired of being nothing. I'm not prime. But at least I'm important when I'm next to other numbers."

SSSSSSSSay! I'm important! I'm number one in everyone's book But I guess I'm not important enough to be a prime number?

How many prime numbers would you guess there are?

There are more prime numbers than you can count. A Greek by the name of Euclid (You-klid) said that there is no end to prime numbers. Prime numbers are infinite. The sign for infinity or "without end" looks like an 8 on its side:

What do you call counting numbers greater than 2 that are not prime?

Counting numbers greater than 2 that are not prime numbers are called composite numbers because they are composed of two or more prime numbers that are multiplied together. Here are some examples of **composite numbers**:

 4 is composed of 2 X 2 = 4
 6 is composed of 2 X 3 = 6
 8 is composed of 2 X 2 X 2 = 8
 9 is composed of 3 X 3 = 9
 12 is composed of 2 X 2 X 3 =12

Why do we learn prime numbers?

Prime numbers help you reduce fractions and are also used to make secret codes that are difficult to break or figure out. If you know about prime numbers, then you have numbers that help you with fractions and factors. It's like you have a magnet that helps you pick up nails.

What's an easy way to find all of the prime numbers from 1 to 100?

1	2	3	4	5	6	7	8	9	10
11	12	13	14	15	16	17	18	19	20
21	22	23	24	25	26	27	28	29	30
31	32	33	34	35	36	37	38	39	40
41	42	43	44	45	46	47	48	49	50
51	52	53	54	55	56	57	58	59	60
61	62	63	64	65	66	67	68	69	70
71	72	73	74	75	76	77	78	79	80
81	82	83	84	85	86	87	88	89	90
91	92	93	94	95	96	97	98	99	100

There's a trick that a Greek by the name of Eratosthenese (air-uh-TOSS-the-knees) thought of around 240 BC call the Sieve of Eratosthenes. A sieve (SEEVE) is a strainer or filter used in cooking.

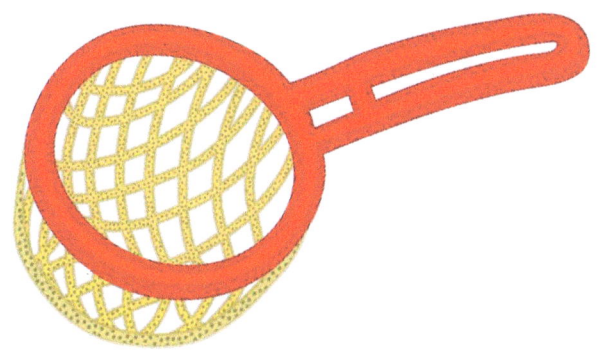

The trick is to have 10 touching boxes in one row (use graph paper or else lined writing paper and draw vertical lines). Each box is the same size and big enough to write the number 100. There are 10 rows with 10 boxes across. Write the numbers across from 1 to 100 each inside a box starting from left to right.

1	2	3	4	5	6	7	8	9	10
11	12	13	14	15	16	17	18	19	20
21	22	23	24	25	26	27	28	29	30
31	32	33	34	35	36	37	38	39	40
41	42	43	44	45	46	47	48	49	50
51	52	53	54	55	56	57	58	59	60
61	62	63	64	65	66	67	68	69	70
71	72	73	74	75	76	77	78	79	80
81	82	83	84	85	86	87	88	89	90
91	92	93	94	95	96	97	98	99	100

Is the number 1 a prime number?

No, the number one is not a prime by definition. A prime number must be a counting (natural) number greater than 1 that can be divided evenly only by 1 and itself without a remainder.

So cross out the number 1.

X	2	3	4	5	6	7	8	9	10
11	12	13	14	15	16	17	18	19	20
21	22	23	24	25	26	27	28	29	30
31	32	33	34	35	36	37	38	39	40
41	42	43	44	45	46	47	48	49	50
51	52	53	54	55	56	57	58	59	60
61	62	63	64	65	66	67	68	69	70
71	72	73	74	75	76	77	78	79	80
81	82	83	84	85	86	87	88	89	90
91	92	93	94	95	96	97	98	99	100

#7. Is the number 2 a prime number?

#7. Yes, the number two is a prime by definition. **A prime number must be a counting (natural) number greater than 1 that can be exactly divided ONLY by 1 and itself** without a remainder. Two can be divided by 1 and itself without a remainder.
So circle the number 2.

✗	②	3	4	5	6	7	8	9	10
11	12	13	14	15	16	17	18	19	20
21	22	23	24	25	26	27	28	29	30
31	32	33	34	35	36	37	38	39	40
41	42	43	44	45	46	47	48	49	50
51	52	53	54	55	56	57	58	59	60
61	62	63	64	65	66	67	68	69	70
71	72	73	74	75	76	77	78	79	80
81	82	83	84	85	86	87	88	89	90
91	92	93	94	95	96	97	98	99	100

Is the number 3 a prime number?

Yes, the number three is a prime by definition. **A prime number must be a counting (natural) number greater than 1 that can be exactly divided <u>ONLY</u> by 1 and itself** without a remainder. Three can be divided by 1 and itself without a remainder.
So circle the number 3.

✗	②	③	4	5	6	7	8	9	10
11	12	13	14	15	16	17	18	19	20
21	22	23	24	25	26	27	28	29	30
31	32	33	34	35	36	37	38	39	40
41	42	43	44	45	46	47	48	49	50
51	52	53	54	55	56	57	58	59	60
61	62	63	64	65	66	67	68	69	70
71	72	73	74	75	76	77	78	79	80
81	82	83	84	85	86	87	88	89	90
91	92	93	94	95	96	97	98	99	100

Is the number 4 a prime number?

No, the number four is a composite number by definition. A composite number is a counting (natural) number greater than 1 that can be exactly divided ONLY by 1 and one or more prime numbers without a remainder. Four can be divided evenly by 2 which is a prime number without any remainder. So, cross out the number 4.

✗	②	③	✗	5	6	7	8	9	10
11	12	13	14	15	16	17	18	19	20
21	22	23	24	25	26	27	28	29	30
31	32	33	34	35	36	37	38	39	40
41	42	43	44	45	46	47	48	49	50
51	52	53	54	55	56	57	58	59	60
61	62	63	64	65	66	67	68	69	70
71	72	73	74	75	76	77	78	79	80
81	82	83	84	85	86	87	88	89	90
91	92	93	94	95	96	97	98	99	100

Is the number 5 a prime number?

Yes, the number five is a prime by definition. **A prime number must be a counting (natural) number greater than 1 that can be exactly divided <u>ONLY</u> by 1 and itself** without a remainder. Five can only be divided by 1 and itself without a remainder. So circle the number 5.

✗	②	③	✗	⑤	6	7	8	9	10
11	12	13	14	15	16	17	18	19	20
21	22	23	24	25	26	27	28	29	30
31	32	33	34	35	36	37	38	39	40
41	42	43	44	45	46	47	48	49	50
51	52	53	54	55	56	57	58	59	60
61	62	63	64	65	66	67	68	69	70
71	72	73	74	75	76	77	78	79	80
81	82	83	84	85	86	87	88	89	90
91	92	93	94	95	96	97	98	99	100

Is the number 6 a prime number?

No, the number six is a composite number by definition. A composite number is a counting (natural) number greater than 1 that can be exactly divided <u>ONLY</u> by 1 and one or more prime numbers without a remainder. Six is composed of only the prime numbers 2 X 3 = 6. So cross out the number 6.

✗	②	③	✗	⑤	✗	7	8	9	10
11	12	13	14	15	16	17	18	19	20
21	22	23	24	25	26	27	28	29	30
31	32	33	34	35	36	37	38	39	40
41	42	43	44	45	46	47	48	49	50
51	52	53	54	55	56	57	58	59	60
61	62	63	64	65	66	67	68	69	70
71	72	73	74	75	76	77	78	79	80
81	82	83	84	85	86	87	88	89	90
91	92	93	94	95	96	97	98	99	100

Is the number 7 a prime number?

Yes, the number seven is a prime by definition. **A prime number must be a counting (natural) number greater than 1 that can be exactly divided <u>ONLY</u> by 1 and itself** without a remainder. Seven can only be divided by 1 and itself without a remainder.
So circle the number 7.

✗	②	③	✗	⑤	✗	⑦	8	9	10
11	12	13	14	15	16	17	18	19	20
21	22	23	24	25	26	27	28	29	30
31	32	33	34	35	36	37	38	39	40
41	42	43	44	45	46	47	48	49	50
51	52	53	54	55	56	57	58	59	60
61	62	63	64	65	66	67	68	69	70
71	72	73	74	75	76	77	78	79	80
81	82	83	84	85	86	87	88	89	90
91	92	93	94	95	96	97	98	99	100

Can you see the composites of the prime number two?

The rule for divisibility by two lets you cross out the multiples of two which end in 2,4,6,8,0. So cross out all the multiples of the prime number two which are composites.

❌	②	③	❌	⑤	❌	⑦	❌	9	❌
11	12	13	14	❌	16	17	18	19	20
❌	22	23	24	25	26	❌	28	29	30
31	32	33	34	35	36	37	38	39	40
41	42	43	44	45	46	47	48	49	50
51	52	53	54	55	56	57	58	59	60
61	62	63	64	65	66	67	68	69	70
71	72	73	74	75	76	77	78	79	80
81	82	83	84	85	86	87	88	89	90
91	92	93	94	95	96	97	98	99	100

Can you see the composites of the prime number three?

The rule for divisibility by three lets you cross out the multiples of three in which the sum of all the digits in the number added together makes 3, 6, or 9. So cross out all the multiples of the prime number three which are composites.

Cross out these composites
 9 because   9 = 9
12 because 1 + 2 = 3
15 because 1 + 5 = 6
18 because 1 + 8 = 9
21 because 2 + 1 = 3
57 because 5 + 7 = 12 and 1+2=3
87 because 8 + 7 = 15 and 1+5=6
and so on to 99.

Note: Even numbers are already crossed out by the prime number two.

Can you see the composites of the prime number five?

The rule for divisibility by five lets you cross out the multiples of five in which the number ends in zero or five. So cross out all the multiples of the prime number five which are composites.

Cross out these composites
10 because it ends in zero
15 because it ends in five
20 because it ends in zero
25 because it ends in five
30 because it ends in zero
35 because it ends in five

Note: Even numbers are already crossed out by the prime number two.

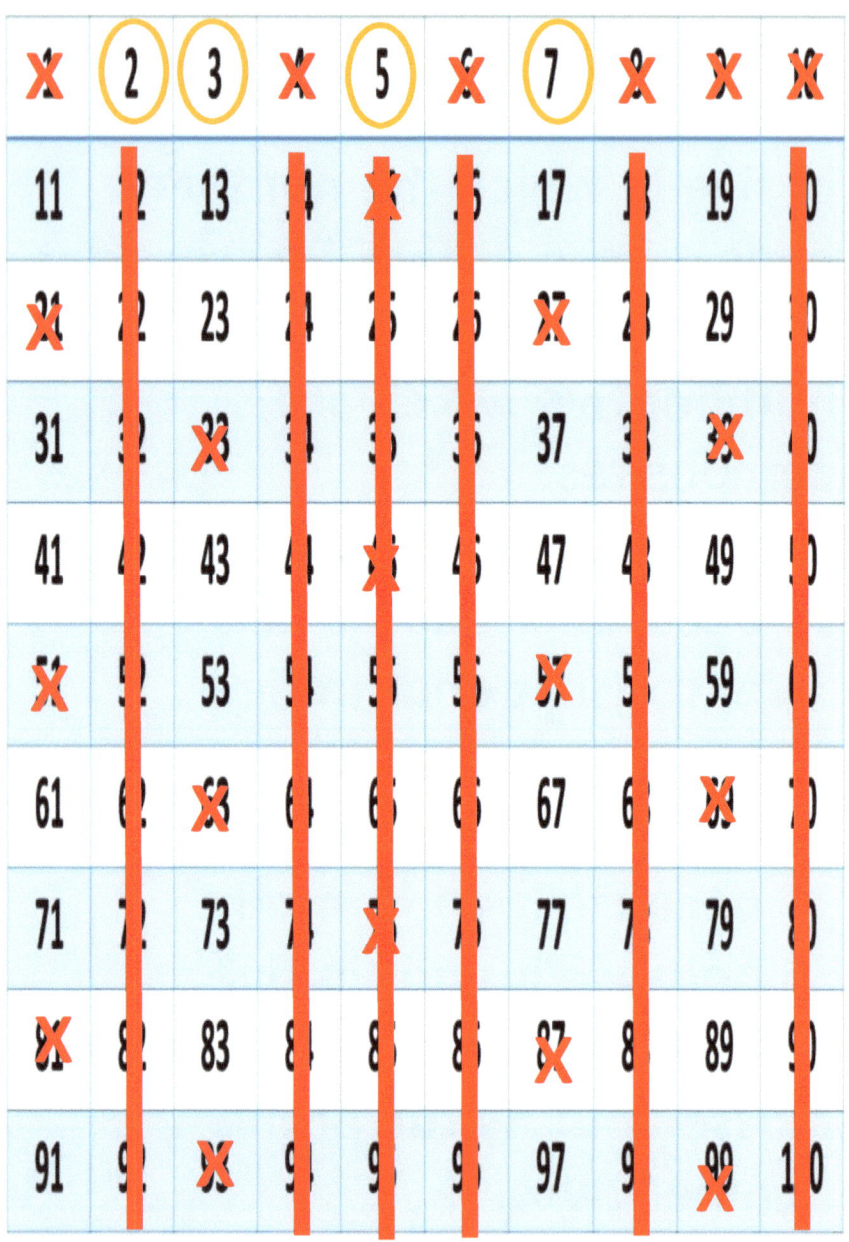

Can you see the composites of the prime number seven?

The rule for divisibility by seven lets you cross out the multiples of sevens. It's easiest to just divide by seven to see if you get no remainder. The multiples of seven up to 100 are: 14, 21, 28, 35, 42, 49, 56, 63, 70, 77, 84, 91, 98

Cross out these multiples of seven.

Note: Some numbers are already crossed out by the prime numbers two, three and five.

I'll pretend bet you a quarter that you can't guess how many prime numbers there are between 1 and 100?

There are 25 prime numbers between 1 and 100. Did you win the 25 cent imaginary bet?

Around 2,000 BC, the Ancient Egyptians thought prime numbers were magical because they help you do math problems.

Around 300 BC, the Ancient Greeks and Ancient Romans also thought prime numbers were mystical. Because they were helpful in solving fractions and other math problems.

Today, primes are used in making secret codes (encryptions) plus credit card and cell phone security. We are still using and learning new things about prime numbers.

There is a Trick for Prime Numbers That Does Not Always Work

Some prime numbers can be written as
(6 X N + 1) or (6 X N-1).
For example, all of these are primes with some exceptions:

6 X 1 - 1 = 5 is prime
6 X 1 + 1 = 7
6 X 2 - 1 = 11
6 X 2 + 1 = 13
6 X 3 – 1 = 17
6 X 3 + 1 = 19
6 X 4 – 1 = 23
6 X 4 + 1 = 25 not a prime
6 X 5 – 1 = 29
6 X 5 + 1 = 31
6 X 6 – 1 = 35 not a prime
6 X 6 + 1 = 37

A Trick for Prime Numbers That Does Not Always Work

Some prime numbers can be written as
(6 X N + 1) or (6 X N-1).
For example, all of these are primes with some exceptions:

6 X 7 - 1 = 41 is prime
6 X 7 + 1 = 43
6 X 8 - 1 = 47
6 X 8 + 1 = 49 not a prime
6 X 9 – 1 = 53
6 X 9 + 1 = 55 not a prime
6 X 10 – 1 = 59
6 X 10 + 1 = 61
6 X 11 – 1 = 65 not a prime
6 X 11 + 1 = 67
6 X 12 – 1 = 71
6 X 12 + 1 = 73

6 X 13 - 1 = 77 not a prime
6 X 13 + 1 = 79
6 X 14 - 1 = 83
6 X 14 + 1 = 85 not a prime
6 X 15 – 1 = 89
6 X 15 + 1 = 91
6 X 16 – 1 = 95 not a prime
6 X 16 + 1 = 97

If you use this trick up to 100 and the answer is 49 or ends in 5 or has digits that repeat themselves like 77 then they are not prime.

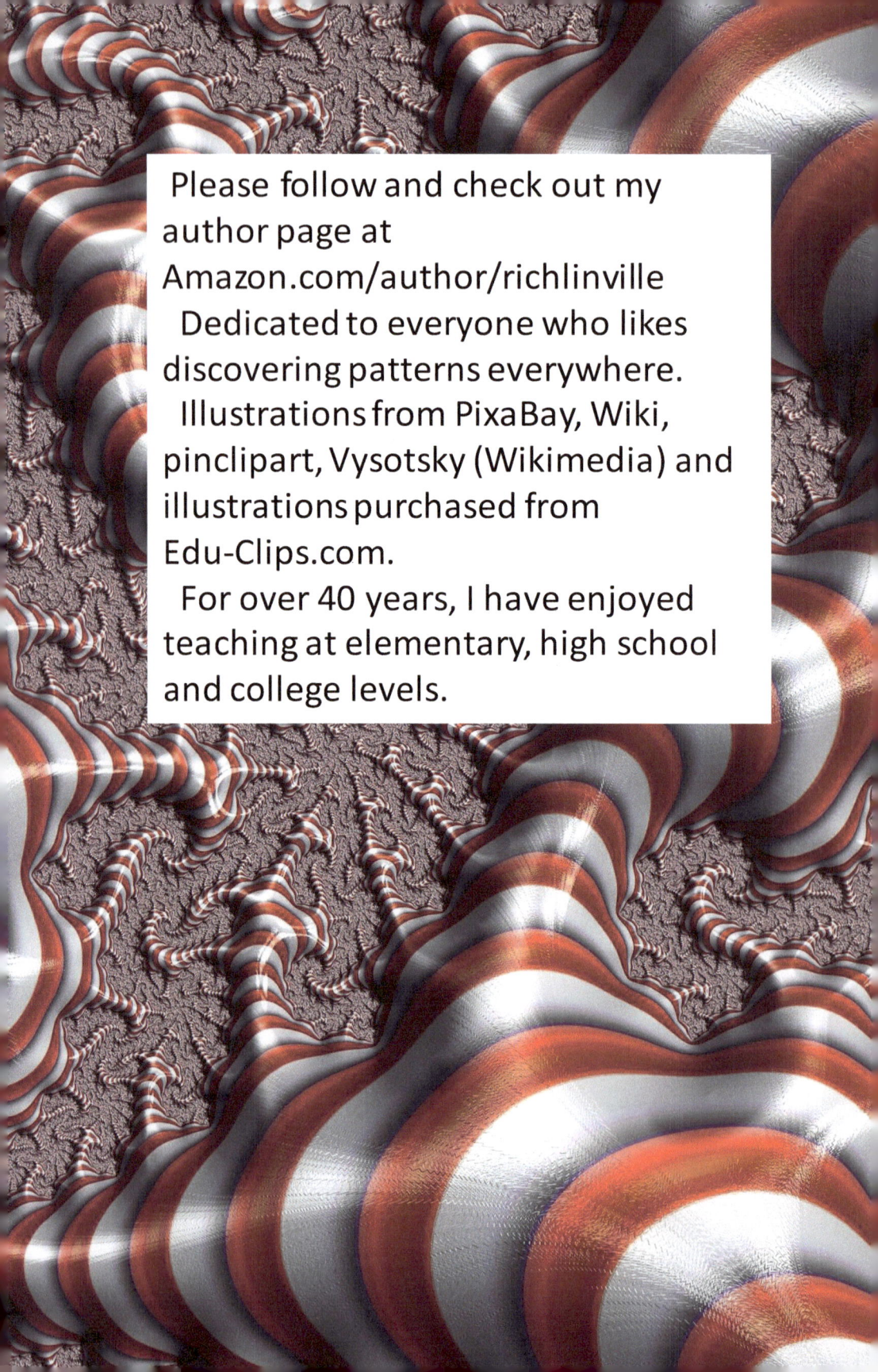

Please follow and check out my author page at Amazon.com/author/richlinville

Dedicated to everyone who likes discovering patterns everywhere.

Illustrations from PixaBay, Wiki, pinclipart, Vysotsky (Wikimedia) and illustrations purchased from Edu-Clips.com.

For over 40 years, I have enjoyed teaching at elementary, high school and college levels.

ISBN: 9798373664981

www.ingramcontent.com/pod-product-compliance
Lightning Source LLC
Chambersburg PA
CBHW040222220526
45473CB00001B/86